COUNTDOWN
TO EXTINCTION

ADAPTED BY Barbara Gaines Winkelman

New York

COUNTDOWN
TO EXTINCTION

CHAPTER ONE

THE PERFECT DINOSAUR

Uncle Grant did not want me to stay with him that summer. All I was to him was a distraction. He is a paleontologist. He studies dinosaurs. And he was obsessed. When he wasn't talking about dinosaurs, I could tell he was thinking about them.

But let me start from the beginning.

My name is Will. My parents are paleontologists like Uncle Grant. That summer Mom and Dad went on a dinosaur dig in Asia, and they thought I'd have fun staying at Uncle Grant's in Florida. Uncle Grant

worked at the Dino Institute down there.

The institute had just opened what they called "a natural history museum of the future." Special cars named CTX Time Rovers took people far back in time. Uncle Grant told me that tachyon particles, which travel faster than light, enable the Time Rovers to travel through time. That's all he would tell me. The rest is a trade secret.

I was trying to complete my own secret mission: to get Uncle Grant to notice me. He barely spoke to me.

One day Uncle Grant came home from work so excited that he actually confided in me—about dinosaurs, of course. He told me that he had been to the late Cretaceous Period and found the perfect subject for further study: an iguanodon.

"Just think," he mused, "iguano bones were the first dinosaur bones ever discovered—over 150 years ago. It's only fitting that an iguanodon would be the first *living* dinosaur to be studied." Suddenly Uncle Grant became very serious. "Just between you and me, Will, I've got to get this iguanodon back to the present."

This got me thinking. Maybe this dinosaur is the key to getting closer to Uncle Grant. I could help Uncle Grant bring it back to our time. We could both take care of it. Then Uncle Grant would like me.

CHAPTER TWO

AN IDEA

Early the next morning the sound of rain woke me up. I got dressed and went to the kitchen for breakfast.

I was thinking about this iguanodon when Uncle Grant walked in and opened the paper, completely ignoring me.

I tried to get his attention. "Uncle Grant, I'd like to get a good look at your iguanodon," I said.

"Sure, sure," said Uncle Grant absently.

I heard the rain beat down harder as I cleared my throat and said, "How about today? Why don't you take me to work with you?"

Uncle Grant paused for a minute and then answered, "Uh, okay. Why not?"

"Great. Can I take Marissa? She loves anything prehistoric," I explained.

Marissa lives next door to Uncle Grant. She is almost as obsessed with prehistoric times as everyone in my family is.

"Yes, that's fine. As long as her mom says it's okay."

Marissa's mom said yes, so the three of us headed to the Dino Institute.

Uncle Grant told us how the Time Rovers worked, during the drive to the museum.

"There are special silicon cells implanted in the Time Rovers," he explained. "The cells read radio waves that bounce from surrounding objects. This allows technicians monitoring the Rovers to identify forms and

take control of the steering when necessary.

"As the tracking system recognizes different dinosaurs, it tells what they are. If the dinosaur is a herbivore, a plant eater, the Rover keeps going. But if the dinosaur is a carnivore, a meat eater, the technicians can steer clear of danger."

Finally we pulled into the museum's parking lot.

"You'll go on one of the morning tours of the Cretaceous Period," Uncle Grant told us.

Uncle Grant led us to the auditorium. "A tour has just started. You can join it or go on

the next tour. I'll see you later. Have fun!"
Uncle Grant called after us as we entered the
auditorium.

We decided to join the tour that had just
started.

First, we had to watch a video of a woman
I recognized: Dr. Helen Marsh, Uncle Grant's
boss.

That's when I told Marissa my idea.

"Hey," I whispered, "maybe the tracking
system can find my uncle's iguanodon and
help us bring it back to the present."

"Don't be crazy!" answered Marissa. "An
iguanodon in this century? What's he going
to do with it?"

"Never mind
that, Marissa. Uncle
Grant and I will
figure that out
together."

CHAPTER THREE

DR. GRANT
SEEKER

There was no more time for talk. Dr. Marsh said, "Now, as the camera takes you live to the control center, I present to you one of our highly skilled staff members, Dr. Grant Seeker, who will give you a complete safety briefing."

There was Uncle Grant on the screen. He was at his desk.

After explaining the safety tips, Uncle Grant began to unfold his plan. "Let's talk about you, and how you can help me make history today," he told the tour group. "It's like this—if I can bring you back from the Cretaceous Period, it stands to reason I can bring a live dinosaur back with you. And not just any dinosaur . . ."

Marissa and I looked at each other.

"My idea!" I said. "Great minds do think alike!"

"Warped minds, maybe," answered Marissa.

A picture of an iguanodon flashed on the screen.

"Take a look at this guy," Uncle Grant said. "He's an iguanodon— one in a million— a real leader. I spotted him during a little unauthorized field trip yesterday and tagged him with a locator."

I could see why Uncle Grant was stuck on

this iguanodon. He was one cool-looking dinosaur. His head was shaped like an egg laid sideways on top of a short neck.

His mouth was wild. It was shaped like a pointed snout, and the bottom lip ended with a birdlike beak!

He stood upright, on weird-looking feet—he had only three toes! His tail stuck out in midair.

His hands were interesting, too. Each had four flexible fingers, but the fifth finger, called a thumb spike, was all nail.

Uncle Grant pointed the camera at a radar screen, where we saw a blinking white dot.

"Right now," explained Uncle Grant, "this iguanodon is in the very late Cretaceous Period. That's where you'll be going today."

"Will," Marissa said, "even though this iguanodon is really awesome, I don't think it's a good idea to go back and get him."

"Why not?"

"Don't you know your prehistoric trivia?" Marissa impatiently asked. "The Cretaceous Period ended suddenly 65 million years ago. All of a sudden the dinosaurs were gone. It was a mass extinction. Scientists think that their sudden disappearance was caused by the impact of a huge meteorite."

"Yeah, yeah. But this iguanodon is still around," I argued.

"Well, he's about to become extinct. And we could become extinct with him if we're not careful," warned Marissa.

"Uncle Grant wouldn't send us if he didn't think it was safe. He can watch us on the tracking system. Look, Marissa. I've got to get that iguanodon for Uncle Grant. Then he'll finally like me."

Just then Dr. Marsh stormed into the

control room and walked up to Uncle Grant. "I'm just in time, it seems, to correct a little misstatement. The end of the Cretaceous Period is, of course, very close to the time when the impact of a giant asteroid destroyed most life forms on Earth. That's why your time-travel coordinates have been locked on the *early* Cretaceous period."

Marissa shot me an I-told-you-so look.

"And I can assure you," continued Dr. Marsh, "that every Rover's computer program is securely encrypted with a secret code so no one can change the time coordinates. Bon voyage!"

I just stood there as Dr. Marsh left in a huff. I was crushed. There went my plan.

But Uncle Grant was smiling as he faced

the camera. "The Time Rovers are securely locked unless, of course, you happen to have the encryption code—which I do!"

Uncle Grant punched the code into the computer. ACCESS GRANTED flashed on the screen.

"We're in!" Uncle Grant excitedly shouted. "Now, here's the plan. You follow the special homing signal to the iguanodon. Then I'll enlarge the transport field, and *boom*, you're back with one spectacularly beautiful iguanodon. Don't worry about that asteroid. You'll be in and out before it even breaks the atmosphere. Trust me. What could go wrong? See you in 65 million years!"

I turned to Marissa. "Well, this is it. If you want to turn back, do it now. But I'm getting that iguanodon for Uncle Grant."

Marissa quickly replied, "An adventure in the Cretaceous Period sure beats a rainy day in Florida. I'm in."

CHAPTER FOUR

ALL ABOARD!

We went into the Time Rover Station. Marissa and I sat down in Rover Number Thirteen. A woman and a girl sat next to us. The girl looked about our age and was all dressed up.

Marissa and I were getting psyched up to go when these two guys came and sat down behind us. These guys were big. I mean really big. Their arms were the size of tree trunks.

Our Rover left the station. It was then that we saw the Time Tunnel—and a red flashing light near the entrance. A siren sounded. Suddenly Dr. Marsh's voice sounded over the two-way radio. She seemed pretty angry.

"Attention. This is Dr. Marsh. Due to a security breach, your tour is being terminated." But the Rover lurched forward into the Time Tunnel.

"Too late," said Uncle Grant. "We're past the point of no return."

The tunnel revved up like a giant genera-
tor. The sounds of the motor got louder
and louder as all sorts of flashing lights
engulfed us.

CHAPTER FIVE

TUNNEL TIME

The Time Tunnel seemed to go on forever. Lights were flashing all around us. My heart was beating so hard that I thought it would jump right out of my chest!

The woman sitting next to me called out, "Hi. I'm Mrs. Hill. This is my daughter, Stephanie. What are your names?"

Marissa and I looked at each other and shrugged before I spoke up and introduced us.

"I'm Will, and this is Marissa."

What is this, I thought, a tea party?

Whatever it was, the big guys didn't want to be a part of it. They didn't answer Mrs

Hill. They didn't even look her way. Mrs. Hill turned around and slapped the guy behind her and shouted, "Hey, fellas, I'm talking to you. What are your names?"

They shuffled slightly.

"Frank," grunted one.

"Curly," grunted the other.

I looked at the girl, Stephanie. She was looking straight up in the air as if no one existed except herself.

"This is going to be interesting," I told Marissa.

CHAPTER SIX

THE CALM BEFORE THE STORM

We left the Time Tunnel and entered a dense forest lit up by the moon.

"It's beautiful here," gasped Marissa. "These must be the ancestors of today's trees."

Uncle Grant's voice came over the radio.

"Okay, let's shed a little light on the past."

The vehicle's lights came on and revealed a styracosaurus scratching himself against a tree.

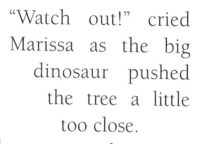

"Watch out!" cried Marissa as the big dinosaur pushed the tree a little too close.

The Rover moved on as small meteors streaked across the sky.

"All those rocks came from space," I said. "I wonder how far they traveled."

"As long as they're small meteors, and not the big one that ended the dinosaurs, we're all right, I guess," Marissa said.

As the Rover climbed over a hill, we stared in disbelief at the forest down below. It was populated by a variety

of dinosaurs. I could recognize dinosaurs from their bones. Marissa knew them all from her books. Now we were both identifying them in the flesh!

"This is awesome!" I exclaimed.

"*Totally* awesome!" agreed Marissa.

"Two minutes to asteroid impact," said a computer voice over the radio.

"Right on schedule," said Uncle Grant.

"Two minutes till a huge meteor hits! Will, forget it. Let's turn around!" Marissa cried. Her eyes widened as she looked behind me. "Yuck!"

Marissa pointed to an alioramus eating a giant lizard. All we saw was the lizard's tail hanging out of the dinosaur's mouth. The tail was still moving.

"Cool!" Stephanie shouted as the alioramus swallowed the last bit of the lizard.

Trying not to stare at Stephanie, I looked to the right and pointed to a smoking mountain. "Is that a volcano?"

Marissa looked at the horizon and replied, "Not only is that a volcano, but it's erupting. Will, this is not such a great idea."

But the ride was too exciting to stop.

"Check out the dino family," I said. There were two babies and a mother, who looked worried.

"Oooooo so cute," cooed Stephanie. "Duckbills! Let's grab the babies and take them back, too."

"They're hadrosaurs," corrected Marissa.

"Don't be so silly, Stephanie," scolded Mrs. Hill.

Thump. The Time Rover was shoved off course by a boulder hitting its side.

"Oh, no!" shrieked Marissa. "We've been hit by one of those meteors!"

The Rover lurched away from the path as sparks flew everywhere. The warning lights flashed on and off.

Over the radio we could hear static, mixed in with Uncle Grant's voice. He seemed to be telling us that the circuits had broken, and he had to put us on autopilot while he fixed them.

I wondered if the computer could steer the Rover when the Rover was off-course.

CHAPTER SEVEN

DINOSAURS UP CLOSE AND PERSONAL

The Rover lurched over the top of a steep hill, and started going down fast.

"Whoa, roller-coaster ride, prehistoric-style!" I shouted.

"EEEEE!" was all Marissa could say. I noticed a smile on her face.

We entered a part of the forest that had no trails.

BOOM! We slammed into something and

jolted to a stop. An animal roared so loudly that it shook the Rover. Flash! A meteor shower turned night into day, and we were face-to-face with a carnotaurus!

This monster was mean. It had scaly skin and two horns on its large head. This huge meat eater was staring at us with large, hungry eyes.

Thankfully, Uncle Grant was back.

"Okay, we're back on-line, and *that* is a carnotaurus!"

The beast screwed up his nasty face, curled his lips, and spit right at us!

"Slime!" Stephanie laughed gleefully.

Curly shouted, "Get this sticky stuff off of me, you—"

He stopped short as the dino lunged forward. His roars shook the Rover. All at once the meteor shower stopped. In the pitch-black, Marissa and I held on to each other for dear life.

We could hear the dino licking his chops. He was getting ready for his next meal—us!

"Dr. Seeker!" Marissa yelled. "Get us out of here!"

"Hurry!" I added.

"Hold on!" Uncle Grant shouted.

I buried my face in the back of my seat.

"MaMaMaMaMaMaMa!" I heard Frank cry.

"Help!" screamed Curly.

Stephanie was laughing wildly.

With a jerk, the vehicle peeled out from under the giant jaws.

Lightning struck the ground, briefly lighting up the sky, and we caught a glimpse of the carnotaurus thundering

through the brush alongside us.

The Time Rover swerved to the left and right as Uncle Grant gained control of the steering. He had managed to get us out of danger—for the time being.

Marissa and I sat back.

"Oh, thank goodness!" Mrs. Hill exclaimed.

We all let out huge sighs of relief.

CHAPTER EIGHT

CALM AGAIN

Whew! That was close," exclaimed Marissa. "That carnotaurus was out for fresh meat."

"Glad to see you're still in good humor," said Uncle Grant over the radio. "That means you're all right."

In the darkness the Rover began to bounce up and down across some rocks.

Suddenly, it became very noisy. Giant insects were buzzing around us. They were much larger than anything in my backyard.

"Wow! Look over there!" Marissa exclaimed. "There's a herd of hadrosaurs."

In the distance I heard a familiar roar

"Oh, no!" I said. "I hear the carnotaurus!"

"His cries are pretty low. He must be way behind us," Marissa said hopefully.

The radio sounded. "All clear and on target," Uncle Grant stated. "You're coming up on a sauropod. Watch out for his tail."

"Squish, squash," shouted Stephanie as we rode over the tail.

"Ouch! That's gotta hurt!" Uncle Grant commented.

We continued forward. We were startled by loud groaning and growling. We braced ourselves for another encounter with a mean giant. A meteor lit up the sky to reveal a huge . . .

"Saltasaurus!" Marissa and I shouted together in relief.

"Good thing he's a vegetarian!" Uncle Grant exclaimed.

I looked over at Frank and Curly. They looked a little confused.

"It's okay," I explained. "This is a kind of sauropod, a saltasaurus. Sauropods are gentle herbivores. Even though they are some of the largest animals that ever lived, we're safe."

"Thanks, Einstein. Now how about getting us out of here?" muttered Frank.

"What a geek!" added Curly.

"Stop this, right now!" scolded Mrs. Hill. "We need to get along if we're going to get through this. Will, thank you for calming us."

But I was too busy studying this

huge animal to care. I had seen sauropod bones a lot—they're pretty common. This one was exactly as I had pictured a saltasaurus. He was gigantic, with thick, elephantlike legs supporting a body that must have weighed as much as *ten* elephants. His front legs were longer than his back legs. His neck was long, much longer than a giraffe's, so he could eat the very highest leaves.

The saltasaurus bent his long neck and sniffed the Time Rover as we moved on.

CHAPTER NINE

SYSTEM ALERT

Ninety seconds to asteroid impact," called out the computer voice.

"Will, the meteorite's coming in ninety seconds!" Marissa repeated.

I leaned toward the radio and pleaded, "Uncle Grant, we can do it. I know it. We'll get that iguanodon for you."

"Kid, you better not get us hurt for a dumb animal," Frank shouted.

Suddenly the car spun out of control. We were whirling around, with sound effects provided by Stephanie.

"*Eeeeeeeeeee!*" she screamed, and then erupted into laughter. We were on another prehistoric roller-coaster ride.

"Watch out!" Marissa screamed as the Rover swerved to miss a family of pterodactyls.

"Oooooooh! Babies!" Stephanie sang out.

We were plowing through overgrown brush. A gigantic birdlike creature swooped down from the blackness overhead.

"Mom at twelve o'clock!" Uncle Grant shouted as we ducked.

A meteor illuminated the protective mother pterodactyl. Her wings spread to their full span. She let out an enraged squeak.

"Here, Tweety!" uttered Stephanie.

This was no Tweety Bird. Her wing span must have been at least six feet. Luckily we were moving quickly away.

Suddenly the ground gave way, and we bounced down a steep, rocky incline. Now everyone was screaming. Just as suddenly, we stopped. The tires spun into the ground, spitting gravel. We were stuck!

The computer stated, "One minute to asteroid impact."

There was an uncomfortable stillness in the air. All was dark. Then we heard the growing sound of many feet running.

The footfalls got louder and louder. Out of the darkness leaped eight or nine dinosaurs as small as chickens.

A meteor lit up the scene as we moved forward. A whole herd of these tiny dinosaurs jumped in front of us.

"Wow!" Marissa and I shouted.

SPLAT—a stray dinosaur got caught by the Rover's left tire.

"Prehistoric roadkill!" Stephanie shouted.

Marissa screamed, "Let's get the iguanodon and get out of here!"

CHAPTER TEN

ESCAPE FROM EXTINCTION

The Rover whipped around the corner at high speed and plunged into total darkness. More meteors provided our light—and showed us that the carnotaurus had caught up with us. He looked as hungry as ever!

BAM! The carnotaurus charged and hit the Rover.

Now we were all screaming.

We spun out, escaping the monster, for the time being.

Through the meteor flashes, we saw his snapping teeth as he tore toward us.

The light from the meteor died out. We could no longer see the carnotaurus, but we still heard his thundering footsteps. The Time Rover stopped abruptly.

Uncle Grant was back. "We've got a power drain. I'm going for the backup unit."

The Rover was at a standstill. We all felt helpless.

GRRRRRRRRR. We looked towards the sound as we felt the carnotaurus's hot breath. His head was right above us! He roared and threw his head back. He was ready to claim his dinner.

Marissa and I sat frozen. Scared stiff. But Marissa snapped out of it.

"Watch out!" she shouted. A meteorite hit the ground right next to us.

Uncle Grant came back just in time. "Okay, we're juiced!" he shouted.

The ground shook from the impact of the meteorite, causing an earthquake—directly underneath us!

Uncle Grant steered the Rover into a forest clearing, and the seismic activity calmed down.

Dare we sit back? Yes. Dare we look up? Yes.

There before us was the most beautiful iguanodon we could ever imagine.

"Jackpot!" exclaimed Uncle Grant. "That's our dino!"

The great dinosaur was near a tree trunk that had fallen. The beast pushed the tree out of the way.

The areas around our Rover and the iguanodon began to sparkle.

Marissa pointed to the sky. "Look up! It's the big one—the asteroid is right up there!"

The sky lit up with a blinding white flash.

"5 . . . 4 . . . 3 . . . 2 . . . 1," stated the computer.

The Rover shuddered from the impact. The white sky became yellow, then orange, and finally red.

Although the meteorite hit far away, we could hear the impact shocks coming toward us. It started as a soft roar, and it got louder and louder until the force of the impact created a vacuum.

It became very windy. Everything was being sucked forward. The Time Rover rocked violently.

I covered my eyes with my hands. I was so scared.

Luckily, the Time Rover finally kicked in.

"Beam me up, Scotty!" shouted Stephanie.

We were scooped up through what seemed to be a hole in space that opened just

for us. Back through the Time Tunnel we went. All sorts of flashing lights surrounded us in the dark.

"Now we can relax," said Mrs. Hill.

"Mom, that was wonderful!" exclaimed Stephanie.

The Time Rover stopped back at the station. We were in the present.

CHAPTER ELEVEN

BACK HOME

Uncle Grant ran toward us, shouting with joy, "Will! Marissa! You're okay!"

"You were both so brave," he added.

Uncle Grant couldn't stop hugging us.

Suddenly, Marissa stepped back and shouted, "Will, Dr. Seeker, look!" She was pointing to a security monitor. We saw the iguanodon running through the halls of the Dino Institute.

"We have to get the dino back!" I said.

Uncle Grant stood still and said, "No, Will. Let security deal with this. Right now, I want to spend more time with my favorite nephew." Uncle Grant put his arm around me.

The iguanodon was caught. He's kept outside the institute as Uncle Grant's and my special project. We take care of him while making amazing discoveries about dinosaurs.

My uncle and I are now as close as we can be.

That's my story. I hope you learned something from it. Always remember: when entering a natural history museum of the future, stay away from the late Cretaceous Period at all cost.

A New World of Adventure Opens at Walt Disney World

Visit Disney's newest theme park, Disney's Animal Kingdom, for an adventure like no other. Go on a heart-pounding expedition into the world of animals on The Kilimanjaro Safari. And in Dinoland, USA, go back 65 million years and witness the end of the dinosaur era—get right in on the action at the thrilling new attraction, Countdown to Extinction. It's all new and it's all amazing.

You can experience the excitement up close with these Disney Press books.

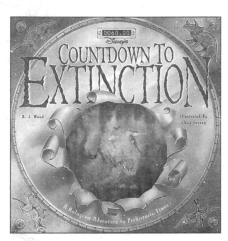

**COUNTDOWN TO EXTINCTION HOLOGRAM BOOK
A HOLOGRAPHIC ADVENTURE
TO PREHISTORIC TIMES**
ISBN 0-7868-3175-8
$16.95 ($22.95 CAN)

**DISNEY CHAPTERS:
DISNEY'S ANIMAL KINGDOM
COUNTDOWN TO EXTINCTION**
ISBN 0-7868-4235-0
$3.95 ($5.50 CAN)

Available at your local bookstore.